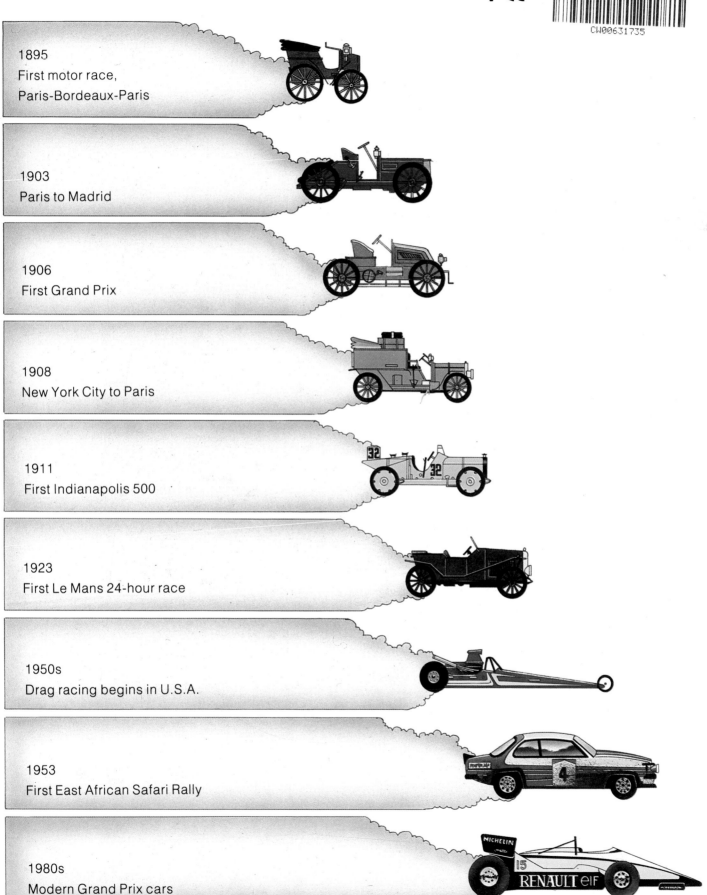

1895
First motor race,
Paris-Bordeaux-Paris

1903
Paris to Madrid

1906
First Grand Prix

1908
New York City to Paris

1911
First Indianapolis 500

1923
First Le Mans 24-hour race

1950s
Drag racing begins in U.S.A.

1953
First East African Safari Rally

1980s
Modern Grand Prix cars

Illustrators
Geraint Derbyshire (John Martin and Artists
Ltd) pages 8–9
Brian Knight (John Martin and Artists Ltd)
pages 12–13
Andy Miles end sheets
Mike Mills pages 10–11
Tom Stimpson cover, pages 6–7, 16–17,
18–19, 20–21, 22–23, 24–25, 26–27, 28–29
Peter Visscher pages 14–15

Editor   Sue Tarsky
Designers David Bennett
           Matthew Lilly

Artists' reference
The National Motor Museum, Beaulieu

First published 1984 by Walker Books Ltd
184–192 Drummond Street
London NW1 3HP

© 1984 Walker Books Ltd

Printed and bound in Spain
by Artes Gráficas Toledo, S.A.
DL-TO-131-84

British Library Cataloguing in Publication Data
Lane, Andrew
    Kings of speed.—(The Story of the car; 2)
    1. Automobiles—Juvenile literature
    I. Title   II. Series
    629.2'222   TL147

ISBN 0-7445-0117-2

CONTENTS

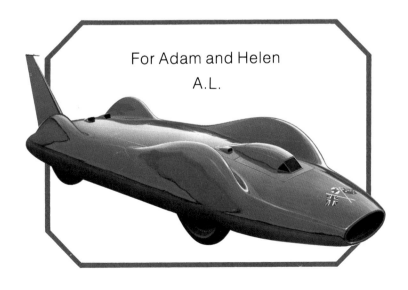

For Adam and Helen
A.L.

KINGS OF SPEED

Written by Andrew Lane

WALKER BOOKS

LONDON

On your mark

1895 Panhard-Levassor

Ever since cars were invented, drivers have raced to see whose car is fastest. The first organised motor race went from Paris to Bordeaux and back, on ordinary roads. Crowds gathered to see the 22 cars start. Like the other drivers, Emile Levassor was alone in his car. Soon he was out in front. When he got to the town where a new driver was to take over, Levassor found him asleep.

Levassor kept on driving. He reached Bordeaux the next day, had a short rest and started back. Night-driving was tiring because it was difficult to see – oil and candle headlamps were very dim.

Levassor was so tired that he fell asleep at the wheel and the car tumbled into a ditch. He pushed the car back on the road and went on to win the race – 48 hours after he had started.

The terrible race

Soon, long-distance races were held between countries. Thousands of people cheered each car as it started the race from Paris, France to Madrid in Spain.

Dust flew up from the unpaved roads and blinded the drivers. They swerved and crashed, and many people were injured. Marcel Renault, a founder of the famous car company, was killed. The race was stopped, and no more races were held on roads open to the public.

The large, powerful racing cars were difficult to control. People stood in the road for a better view; dogs and little children ran in front of the cars.

1903 Mercedes

1903 Renault

Round the world in 168 days

On a cold day in February 1908, one American car, one Italian car, one German car and three French cars left New York City for Paris in the world's greatest race. Each car was packed with food, tools and spare parts. The French De Dion Bouton carried a mast and sails, skis for the wheels and special wheels made to fit on railway lines.

The Italian team, in a Zust, was chased by wolves.

Only the Thomas Flyer, German Protos and the Zust reached the west coast. They went by ship from there to Russia.

There were no roads in Russia – just floods, thick mud and snow.

There were only rough tracks. Soon all the cars were stuck in deep snow.

The American Thomas Flyer went on railway lines and was lifted off when it met trains.

The Zust caught fire and its drivers were arrested as spies.

The Protos reached Paris first but had a penalty of 15 days, so the Thomas Flyer won when it reached Paris 11 days later.

The first Grand Prix

The first Grand Prix in 1906 was a French race that lasted two days, with the cars locked away overnight. The cars were big and heavy and their brakes were not very good, so there were many crashes. Wooden fences kept people off the roads.

Some roads were tarred to stop dust from blinding the drivers. But it was so hot during the race that the tar melted and splashed the men's faces. Even though they wore leather caps and masks with goggles, many drivers were burned.

1906 Clement Bayard

1906 Renault 3A

Grand Prix today

Grand Prix racing has changed since that first race. There are now 16 of these races held round the world. Each lasts two hours. In Monaco the race is held through the streets; in Britain it is held at Silverstone and Brands Hatch. Cars are lightweight, streamlined and low. They are specially built by teams such as Renault.

We're here at last. My mechanics and manager unload the car, tyres and parts.

I'm wearing fire-proof clothes and my helmet has a tube to an oxygen tank, in case of fire. I'm ready.

We're off! I can see the crowd cheering, but all I can hear is the roar of engines.

Raining now. Had to make this pit stop to change to tyres with tread. Well done, team. Just 13 seconds – won't slow me down too much.

We go straight to the pit to get the car ready for the race.

I can see the grand stand filling up with people. Everyone is excited.

Coming out of the straight part — have to start slowing down for the curve.

Oh! Someone has hit the safety barrier! The marshals are there, so he'll be OK. Careful!

The flag is down! I've done it! I'll get a cup, and points for winning. Add them to my other points — I may be world champion this year!

The Count vs. the Red Devil

One chilly December morning in 1898 a small, box-like car raced at full speed through a park near Paris. It was timed at 39 miles per hour. The first World Land Speed Record was set. The electric powered 'Jeantaud' was driven by Count de Chasseloup-Laubat, who was delighted by his amazing triumph.

A Belgian racing driver, Camille Jenatzy, challenged the Count. Jenatzy, known as the 'Red Devil' because of his red hair, drove 'La Jamais Contente', a cigar-shaped electric car. The men met in the park and took turns to see who could drive faster. A few weeks later the Red Devil beat the Count.

The golden age

To try for the Land Speed Record today, a car is timed through a measured mile. It must turn round, refuel and make a return run. The final speed, kept in miles per hour, is the average of the two runs. Early record attempts were made on ordinary roads and then beaches. Many records were set during the 1920s. The Englishman Sir Henry Segrave drove a car with two huge aircraft engines. He had to drive into the water at Daytona Beach, Florida to slow it down when he set a record.

SUNBEAM

TRIPLEX
by WHITE

The American Ray Keech made the next record in 'White Triplex', the largest car engine ever made. The brakes were poor and it was difficult to drive because of its size, so Keech was lucky to survive his dangerous record run.

8.4m long

Segrave wanted the record again, so the streamlined 'Golden Arrow' was built. Thousands of people went to watch as Segrave captured the record again.

Sir Malcolm Campbell

As speeds got faster, cars used the Bonneville Salt Flats in Utah, U.S.A. – a huge, flat desert plain. In the dry season, Bonneville is the perfect place for record-breaking. Sir Malcolm Campbell, an Englishman, was the greatest record-breaker of all. He held the record nine different times.

8.5m long

2.1m wide

His attempt at going 300 miles per hour in 'Bluebird' was full of danger. The windscreen became coated in oil, so it was difficult for Campbell to see.

Exhaust fumes filled the cockpit, so it was difficult for him to breathe, and a tyre caught fire. But Campbell set a record of 301.13 miles per hour.

Jet and rocket cars

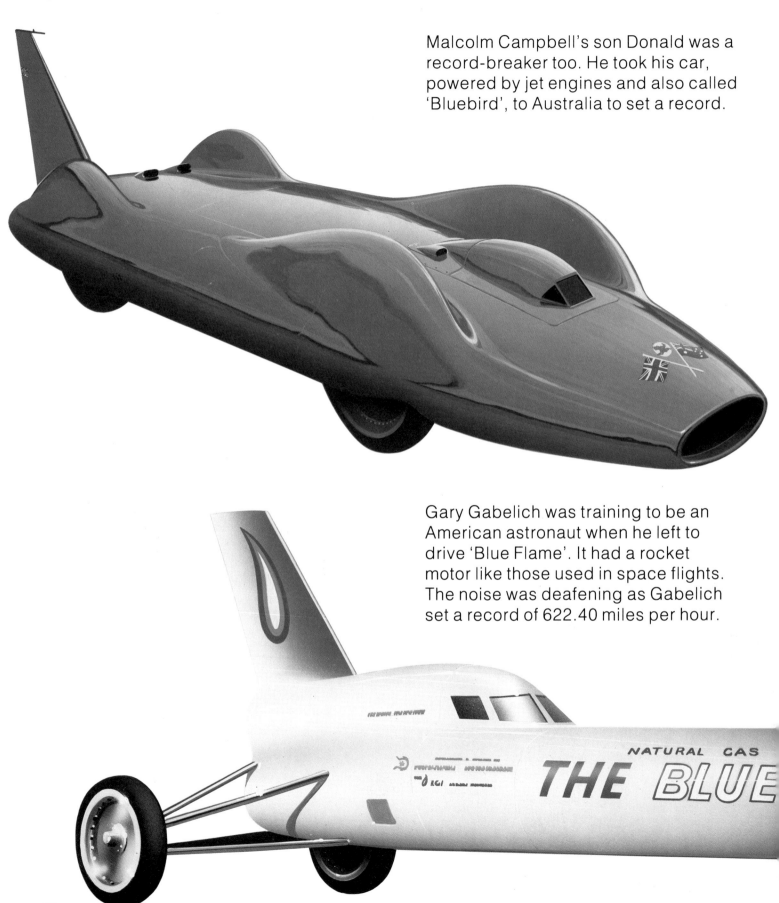

Malcolm Campbell's son Donald was a record-breaker too. He took his car, powered by jet engines and also called 'Bluebird', to Australia to set a record.

Gary Gabelich was training to be an American astronaut when he left to drive 'Blue Flame'. It had a rocket motor like those used in space flights. The noise was deafening as Gabelich set a record of 622.40 miles per hour.

NATURAL GAS
THE BLUE

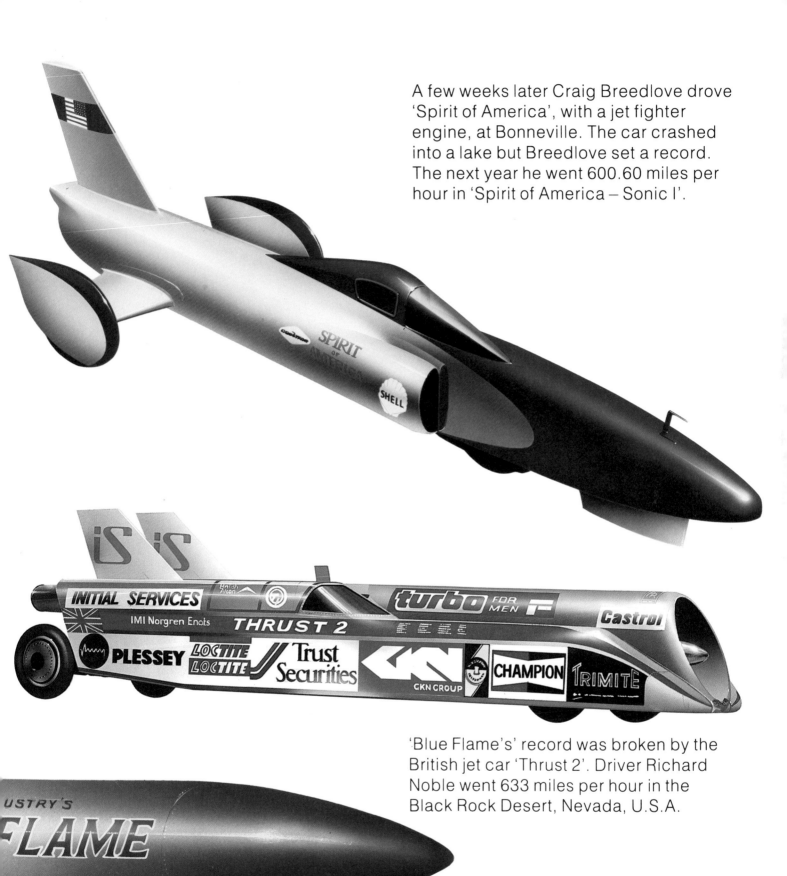

A few weeks later Craig Breedlove drove 'Spirit of America', with a jet fighter engine, at Bonneville. The car crashed into a lake but Breedlove set a record. The next year he went 600.60 miles per hour in 'Spirit of America – Sonic I'.

'Blue Flame's' record was broken by the British jet car 'Thrust 2'. Driver Richard Noble went 633 miles per hour in the Black Rock Desert, Nevada, U.S.A.

Drag racing

Dragster drivers need to have a love of high speed. They can reach 400 km per hour in just five seconds. Dragsters race down a 402-metre strip, alone or two against each other. The fastest have rocket motors and braking parachutes.

Drag racing began in the U.S.A. when people raced cars away from traffic lights. Now it is a popular sport for people to watch. There are two kinds of dragsters. The 'Top Fuel' car has a long, thin body with small front wheels.

The 'Funny Car' looks ordinary, but it is so powerful and lightweight that for a driver, it is like sitting on top of an engine. All drivers wear fire-proof clothes and breathing masks to protect themselves from fire and fumes.

Rallying

Rallying over rough ground at high speeds is a tough test of cars and drivers. Covering long distances, a rally has check points between the start and finish and a set time to drive the course. Navigators guide the drivers with instructions and maps. Cars look ordinary but have powerful engines and strong bodies. The East African Safari is considered the most difficult rally in the world, because the conditions are terrible.

Few of the starters ever finish.

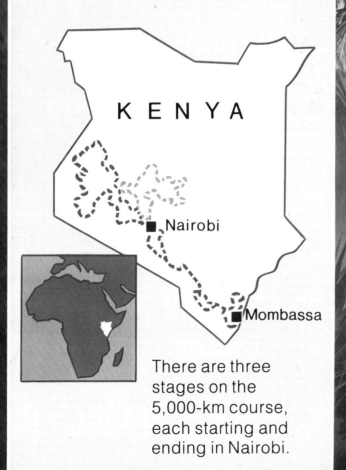

KENYA

■ Nairobi

■ Mombassa

There are three stages on the 5,000-km course, each starting and ending in Nairobi.

Opel Ascona

Wild animals can suddenly run across the bumpy dirt tracks that are full of choking red dust.

Rains turn streams into torrents.

Most cars cannot take the beating.

Audi Quattro

Sports cars

Sports cars, specially made for speed, race on circuits. The toughest is Le Mans in France. Half-a-million people go there every year to watch this 24-hour endurance test of cars and drivers.

There are two drivers for each car so that one can rest while the other races. They swap places every few hours during a pit stop. All night, sometimes in heavy rain, cars roar round the circuit.

There is one driver for each car in the American Indianapolis 500, the world's most famous sports car race. Speeds are faster in this race and the prize money is higher – over one million dollars.

The Indianapolis 500 is like all car racing – exciting and dangerous, with drivers going as fast as they can. Racing drivers and Land Speed Record-breakers are the kings of speed.

Land Speed Record-breaking chart

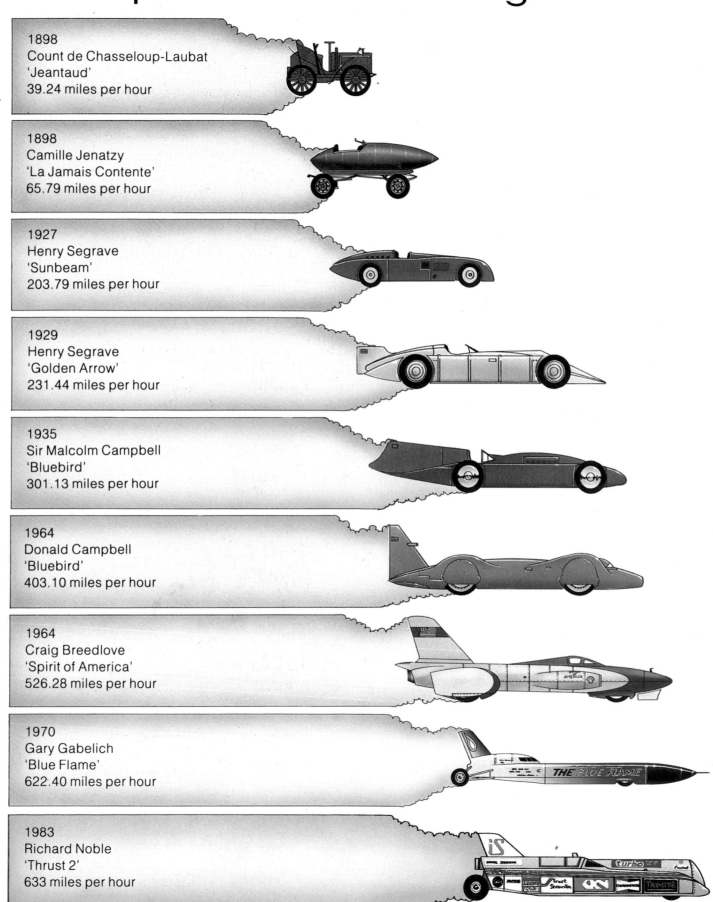

1898
Count de Chasseloup-Laubat
'Jeantaud'
39.24 miles per hour

1898
Camille Jenatzy
'La Jamais Contente'
65.79 miles per hour

1927
Henry Segrave
'Sunbeam'
203.79 miles per hour

1929
Henry Segrave
'Golden Arrow'
231.44 miles per hour

1935
Sir Malcolm Campbell
'Bluebird'
301.13 miles per hour

1964
Donald Campbell
'Bluebird'
403.10 miles per hour

1964
Craig Breedlove
'Spirit of America'
526.28 miles per hour

1970
Gary Gabelich
'Blue Flame'
622.40 miles per hour

1983
Richard Noble
'Thrust 2'
633 miles per hour